Algebra Review Before College

By:

Henry Mark Smith

Introduction

This book is intended to assist students who are about to take a placement test at a college or university, and those students who are about to take a final examination in Intermediate Algebra.

Often, students complete their algebra early, and there is a period of two or three years during which they take advanced mathematics in high school. Some of these students, while skilled in mathematics, forget enough algebra that they are placed in a developmental course. A quick review, not a detailed reading of an algebra textbook, should bring enough algebra back to mind as to eliminate the need for taking a developmental course. This book is such a review, a brief but informative coverage of major topics that are likely to constitute an Intermediate Algebra course, plus a short coverage of functions.

Other students should take the Intermediate Algebra course in entirety. However, these students can still benefit from this book by using it to review for their final examination.

Table of Contents

Chapter 1: The Linear Equation

When an equation is solved, the goal is to isolate the variable on one side of the equation, and its value on the other.

Remember, the symbol = means that the two expressions it joins are equal, so whatever is done to change the value one expression should be done to the other. This means that it is acceptable to add, subtract, multiply, or divide both sides of an equation by the same number, except it is NEVER allowed that anything be divided by zero. It is also not advisable to multiply both sides of an equation by zero, since that effectively removes the information contained in the equation.

In order to solve an equation, the first step should be to remove all symbols of grouping, such as parentheses and fraction bars. This can be done by any legitimate operation, but the distributive law is most frequently used.

$$a(b+c)=ab+ac$$

The arithmetic operations should often be performed in the opposite order as indicated by the order of operations, simply because something, the equation, is being unraveled. So add and/or subtract, then multiply and/or divide.

Example 1.1 Solve the equation $3(x-5)+2x=2(x-6)+9.$

First, eliminate the parentheses, using the distributive law.

$$3x-15+2x=2x-12+9$$

Next, simplify by adding/subtracting like terms on each side of the equation.

$$5x-15=2x-3$$

Add $-2x$ to each side so the variable is eliminated from the right side, and add 15 to each side to eliminate the numerical term from the left side. (The variable could just as correctly be eliminated from the left side, and the numerical term from the eliminated from the right side.)

$$3x = 12$$

Divide both sides of the equation by 3.

$$x = 4$$

Occasionally, the variable will disappear from both sides of the equation simultaneously. When this happens, if the remaining expression is true, all real numbers are solutions. However, if the resulting expression is false, the equation has no solution.

Example 1.2 Solve the equation $5(x+2)-2x+8=3(x+6)$, or state there is no solution.

First, eliminate the parentheses, using the distributive law.

$$5x+10-2x+8=3x+6$$

Next, simplify.

$$3x+18=3x+6$$

Subtract $3x$ from each side.

$$18 = 6$$

False, so there is no solution.

Example 1.3 Solve the equation $4(x-6)+2x=9(x-3)-3(x-1)$, or state there is no solution.

First, eliminate the parentheses, using the distributive law. Observe negative signs are treated as though they are part of the number that follows.

$$4x - 24 + 2x = 9x - 27 - 3x + 3$$

Next, simplify.

$$6x - 24 = 6x - 24$$

Subtract $6x$ from each side.

$$-24 = -24$$

True, so all real numbers are solutions.

Inequalities are handled in the same manner as equations, except the inequality symbols reverse as a result of multiplication or division by a negative number.

Example 1.4 Solve the inequality if it has a solution.

$$-6(x - 4) \leq 3x - 30$$

First, remove all symbols of grouping.

$$-6x + 24 \leq 3x - 30$$

Next, subtract $3x$ and 24 from each side of the inequality.

$$-9x \leq -54$$

Divide each side by -9, reversing the symbol because the divisor is negative.

$$x \geq 6$$

The solutions are all real numbers that are greater than or equal to 6.

Just like for equations, inequalities may have no solution, or may have an infinite number of solutions. If the variable disappears, and the remaining expression is false, there is no solution to the inequality. If the variable disappears, and the remaining expression is true, all real numbers are solutions of the inequality.

Absolute value equations of the form $\left|ax + b\right| = c$ can be solved by assuming the expression between the absolute bars might be negative, or might not. This leads to the two expressions $-\left(ax + b\right) = c$ and $ax + b = c.$ Solving an absolute value equation is the same as solving two linear equations.

Example 1.5: Solve the equation $\left|2x - 6\right| = 12$

$$-\left(2x - 6\right) = 12 \qquad\qquad 2x - 6 = 12 \qquad \text{Write the two equations, then}$$
solve.

$$-2x + 12 = 12$$

$$-2x = 12 - 12 \qquad\qquad 2x = 12 + 6$$

$$-2x = 0 \qquad\qquad\qquad 2x = 18$$

$$x = 0 \qquad\qquad\qquad\quad x = 9$$

The solutions are 0 and 9

Absolute values inequalities require two linear inequalities to be solved. In addition, if the absolute value inequality involves either the symbol < or $\leq$ with the absolute value to the left of the inequality symbol, the solutions to both linear inequalities are required it the absolute value inequality is to be solved. This requires joining the two solution sets with the word *and*, or writing the solutions as an inequality triple. If, however, the absolute value inequality involves the symbol > or $\geq$ with the absolute value to the left of the absolute value symbol, the inequality is solved by a solution to either linear inequality. This requires joining two solution sets with the word *or*, and writing as a triple is not an option.

$$\left|ax + b\right| < c \text{ is solved by}$$

Option 1.

$$-(ax+b)<c \text{ and } ax+b<c,$$

or

Option 2.

$$-c<ax+b<c$$

$|ax+b|>c$ is solved by

$$-(ax+b)>c \text{ or } ax+b>c$$

Example 1.6: Solve $|x-7|<4.$

Option 1.

$-(x-7)<4$ and $x-7<4$ Write as two linear inequalities.

$x-7>-4$ Multiplying the leftmost inequality by $-1.$

$x>3$ and $x<11$

or

Option 2.

$-4<x-7<4$ Treat as two inequalities, so add 7 to each of the three parts.

$3<x<11$

Example 1.7: Solve $|x+6|>8.$

$-(x+6) > 8$ or $\quad x+6 > 8$ Write as two linear inequalities.

$$x+6 < -8$$

$$x < -14 \qquad \text{or} \qquad x > 2$$

An absolute value cannot be negative, so any equation that has an absolute value equal to a negative number has no solution. Absolute value inequalities that have an absolute value *less than* or *less than or equal to* a negative number likewise have no solution, since any number less than or equal to a negative number is also negative. However, absolute values that are *greater than* or *greater than or equal to* a negative value have solutions where the absolute value is greater than or equal to zero. If this is explored carefully, it is found that absolute values *greater than* or *greater than or equal to* a negative value have all real numbers as solutions, except where otherwise they are not defined, such as divisions by zero.

Application problems involving linear equations can take many forms. A few popular types are reviewed below.

Applications involving the formula $d = rt$ abound. Frequently, d is distance, r is rate or speed, and t is time.

Example 1.8: A person jogging around a track has been jogging at an average speed of 3.5 miles per hour for 30 minutes. What is the total distance jogged by the person? First, notice miles per hour and minutes are incompatible. Change the 30 minutes to half an hour.

Apply the formula $d = rt.$

$$d = 3.5\,\frac{\text{mi}}{\text{h}} \times \frac{1}{2}\,\text{h}$$

Cancel h from the numerator and the denominator, and multiply.

$$d = 3.5\,\frac{\text{mi}}{\cancel{\text{h}}} \times \frac{1}{2}\,\cancel{\text{h}} = 1.75\,\text{mi}$$

Another common theme is the mixture problem. This can be a problem involving a solution, or a price – value problem. Here, it is important to remember that before and after mixing you have the same quantity and value. In the example below, the second and third columns need to add to the same total before and after mixing.

Example 1.9: How many pounds of dog food that costs $4.00 per pound should be mixed with 16 pounds of dog food that costs $3.00 per pound to get a mixture that costs $3.25 per pound?

	Price	*Amt*	*Value*
Before	4.00	x	$4x$
	3.00	16	48
After	3.25	$16+x$	$4x+48$

The equation is derived from the *Price* times the *Amount* after mixing equal to the *Value*.

$$3.25(16+x) = 4x+48$$

This is a linear equation, so solve it like any other linear equation.

$$52+3.25x = 4x+48$$

$$4 = 0.75x$$

$$\frac{4}{0.75} = x$$

$$x = \frac{16}{3} \text{ or } 5\tfrac{1}{3}$$

In a mixture problem involving a solution the *Concentration* times the *Amount* equals the *Amount of the Substance*. Both the *Amount* and the *Amount of the Substance* are preserved in mixing.

Interest problems can also be modeled by $d = rt.$

Example 1.10:

John invested $50,000, some at 4% simple interest and some at 6% simple interest. After one year, John has earned $2600 interest. How much did John invest at each rate?

If x is the amount invested at 4%, then $50,000 - x$ is the amount invested at 6%, since both amounts must total 50,000.

$$0.04x + 0.06(50,000 - x) = 2600$$

$$0.04x + 3000 - 0.06x = 2600$$

$$-0.02x = -400$$

$$x = \frac{-400}{-0.02}$$

$$x = 20,000$$

$20,000 is invested at 4% and $30,000 is invested at 6%.

Using geometry is also a possible theme in application problems.

Example 1.11: What are the dimensions of a field that is twice as long as it is wide, if it takes 6000 feet of fence to enclose it?

P=2L+2W

Let $W = x$. Then, $L = 2x$.

$$6000 = 2(2x) + 2x$$

$$6000 = 4x + 2x$$

$$6000 = 6x$$

$$\frac{6000}{6} = x$$

$$1000 = x$$

The width is 1000 feet, so the length is 2000 feet.

Chapter 2: Exponents and Polynomials

Exponent expressions are expected to be simplified, and students are frequently asked to do this.

Rules for Simplifying Exponents:

1. No variable should appear more than once.
2. All exponents are positive.
3. Any number is expresses as a single reduced fraction or as a single decimal.

Simplifying exponents requires using the laws of exponents, frequently more than one law, in a problem. Often, there will be more than one way to simplify an expression containing exponents, and they are equally valid approaches to the same simplification.

The laws of exponents listed below are valid, provided the expressions are defined. That is, unless there is a division by zero, or an even root of a negative number, these laws will hold. The only new condition is that it is not allowed to raise zero to the zero power.

Laws of Exponents

1. $a^n a^m = a^{n+m}$

2. $\left(a^n\right)^m = a^{nm}$

3. $\dfrac{a^n}{a^m} = a^{n-m}$

4. $a^0 = 1$, provided $a \neq 0$

5. $a^{-n} = \dfrac{1}{a^n}$

6. $(ab)^n = a^n b^n$

7. $\left(\dfrac{a}{b}\right)^n = \dfrac{a^n}{b^n}$

While the laws of exponents are formally defined as operating on no more than two expressions, it is possible to extrapolate them to include combining three or more expressions.

Example 2.1: Simplify $\left(5xy^2\right)^{-3}\left(2x^{-8}y^2\right)^2$.

$$\left(5xy^2\right)^{-3}\left(2x^{-8}y^2\right)^2 = 5^{-3}x^{-3}\left(y^2\right)^{-3}2^2\left(x^{-8}\right)^2\left(y^2\right)^2 \qquad \text{Using law 6}$$

$$= 5^{-3}x^{-3}y^{-6}2^2x^{-16}y^4 \qquad \text{Using law 2}$$

$$= 5^{-3}\cdot2^2x^{-19}y^{-2} \qquad \text{Using law 1}$$

$$= \dfrac{2^2}{5^3x^{19}y^2} \qquad \text{Using law 5}$$

$$= \dfrac{4}{125x^{19}y^2}$$

Example 2.2: Simplify $\left(\dfrac{4x}{yz^2}\right)^{-4}\left(\dfrac{8xy^2}{z^{-5}}\right)^3$.

$$\left(\dfrac{4x}{yz^2}\right)^{-4}\left(\dfrac{8xy^2}{z^{-5}}\right)^3 = \dfrac{\left(4x\right)^{-4}\left(8xy^2\right)^3}{\left(yz^2\right)^{-4}\left(z^{-5}\right)^3} \qquad \text{Using law 7}$$

$$= \frac{4^{-4} x^{-4} 8^3 x^3 \left(y^2\right)^3}{y^{-4} \left(z^2\right)^{-4} \left(z^{-5}\right)^3}$$

Using law 6

$$= \frac{4^{-4} x^{-4} 8^3 x^3 y^6}{y^{-4} z^{-8} z^{-15}}$$

Using law 2

$$= \frac{4^{-4} \cdot 8^3 x^{-1} y^6}{y^{-4} z^{-23}}$$

Using law 1

$$= \frac{4^{-4} \cdot 8^3 x^{-1} y^{10}}{z^{-23}}$$

Using law 3

$$= \frac{8^3 y^{10} z^{23}}{4^4 x}$$

Using law 5

$$= 2 x^3 y^8 z^{23}$$

Example 2.3: Simplify $\left(\dfrac{4x^2 y^3}{12z^5}\right)^0$.

$$\left(\frac{4x^2 y^3}{12z^5}\right)^0 = 1$$

Using law 4

     Polynomials are simplified when no variable appears more than once in any term, all exponents are positive integers, all multiplications and divisions are complete, like terms are combined, and all numbers are reduced fractions or decimals.

When adding or subtracting polynomials, add or subtract like terms. Notice not all terms in one polynomial need have a like term in the other polynomial.

Example 2.4: Simplify $(3x^3 + 4x + 5) + (7x^2 - 6x - 2)$

$$(3x^3 + 4x + 5) + (7x^2 - 6x - 2) = 3x^3 + 7x^2 - 2x + 3$$

When subtracting polynomials, distribute the –1, denoted by the negative sign, to all terms in the polynomial being subtracted.

Example 2.5: Simplify: $(5x^2 - 7x + 10) - (2x^2 + 4x + 6)$

$$(5x^2 - 7x + 10) - (2x^2 + 4x + 6)$$

$$= 5x^2 - 7x + 10 - 2x^2 - 4x - 6$$

$$= 3x^2 - 11x + 4$$

Multiplication requires each term of one factor be multiplied by each term of the other factor. This can be easily proven with the distributive property. More importantly, the use of the laws of exponents in performing the multiplications is essential.

Example 2.6: Multiply $(x - 7)(x^2 - 3x + 5)$.

$$(x - 7)(x^2 - 3x + 5)$$

$$= x \cdot x^2 - 3x \cdot x + 5x - 7x^2 - 7(-3x) - 7(5)$$

$$= x^3 - 3x^2 + 5x - 7x^2 + 21x - 35$$

$$= x^3 - 10x^2 + 26x - 35$$

Division also makes use of the laws of exponents, and is best thought of as dividing the lead term into the lead term.

Example 2.7: Divide. $(2x^2 - 8x + 5) \div (2x - 3)$

$$\begin{array}{r} x \\ 2x-3 \overline{\smash{)}\ 2x^2 - 7x + 5} \end{array}$$

Divide $2x$ into $2x^2$.

$$\begin{array}{r} x \\ 2x-3 \overline{\smash{)}\ 2x^2 - 7x + 5} \\ 2x^2 - 3x \end{array}$$

Multiply.

$$\begin{array}{r} x \\ 2x-4 \overline{\smash{)}\ 2x^2 - 7x + 5} \\ \underline{-2x^2 + 3x } \\ -4x \end{array}$$

Subtract by changing the signs and adding.

$$\begin{array}{r} x \\ 2x-4 \overline{\smash{)}\ 2x^2 - 7x + 5} \\ \underline{-2x^2 + 4x } \\ -4x + 5 \end{array}$$

Bring down the next term.

$$\begin{array}{r} x - 2 \\ 2x-4 \overline{\smash{)}\ 2x^2 - 7x + 5} \\ \underline{-2x^2 + 4x } \\ -4x + 5 \end{array}$$

Divide the lead term into the lead term.

16

$$\begin{array}{r} x \;-\; 2 \\ 2x-4\overline{)2x^2 \;-\;7x \;\;+5} \end{array}$$

$$\underline{-2x^2 \;\;+4x}$$
$$-4x\;+5$$
$$-4x\;+8$$

Multiply.

$$\begin{array}{r} x \;-\; 2 \\ 2x-4\overline{)2x^2 \;-\;7x \;\;+5} \end{array}$$

$$\underline{-2x^2 \;\;+4x}$$
$$-4x\;+5$$

Subtract.

$$\underline{+4x\;-8}$$
$$-3$$

The answer is $x - 2 - \dfrac{3}{2x-4}$, or $x - 2 \text{ R} -3$

   If a division involves one or more polynomials that have a missing term of degree less than the highest term, it must be included with a multiplier of zero.

   An example of this would be $\dfrac{x^2 - 5}{x^3 + 2x - 1}$. This would be set up as shown below.

$$x^2 + 0x - 5\overline{)x^3 + 0x^2 + 2x - 1}$$

Chapter 3: Factoring

Factoring is a tool, not a stand alone topic, so it is quite necessary. In essence, when a polynomial is factored it is broken down into simpler polynomials that, when multiplied together, would reproduce the polynomial.

The first form of factoring to be considered is the extraction of a common factor. This is simply an application of the distributive property.

Example 3.1: Factor $3x^3 + 6x^2 + 30x$.

$3x^3 + 6x^2 + 30x = 3x(x^2 + 2x + 10)$ Notice the distributive property is used, and the knowledge of the laws of exponents is necessary.

Using the common factor idea can be expanded to factoring by grouping, where terms are grouped so a common factor can be determined.

Example 3.2: Factor $ax + a + bx + b$.

$ax + a + bx + b = a(x + 1) + b(x + 1)$ The distributive property is applied to parts of the expression.

$a(x + 1) + b(x + 1) = (x + 1)(a + b)$ The common factor $x + 1$ is extracted.

It should be understood that not every grouping leads to a successful factoring, so pair up the terms carefully.

In trinomial factoring the goal is to undo a multiplication of two binomials. The method of factoring $x^2 + ax + b$ is to break b into all possible integer factors, then find the pair that add to a.

Example 3.3: Factor $x^2 - 2x - 35$.

Breaking –35 into factors gives: $-35, 1; -7, 5; -5, 7, -1, 35$. The pair that sums to –2 is –7, 5. So, $x^2 - 2x - 35 = (x - 7)(x + 5)$.

If the lead coefficient is neither 1 nor –1, it also must be broken into its factors, except no sign change is necessary. It is then necessary to multiply one set of factors from the lead coefficient by one set of factors from the constant term, paying attention to order, and sum these products to find the coefficient of the middle term.

Example 3.4: Factor $2x^2 - 7x + 5$.

Breaking the 5 into factors gives: 1, 5 and –1, –5. Breaking the 2 into factors gives: 2, 1. Selecting –1, –5 and 2, 1 the –7 can be found by multiplying the –1 from the first pair by the 1 from the second pair and the –5 from the first pair be the 1 from the second pair.

$2x^2 - 7x + 5 = (2x - 5)(x - 1)$ Undoing polynomial multiplication the middle term comes from the inner pair and the outer pair.

There are several special factoring formulas with which every student should be familiar. If the first three are forgotten, revert to the method above. However, the last two must be committed to memory.

1. $a^2 - b^2 = (a - b)(a + b)$

2. $a^2 + 2ab + b^2 = (a + b)^2$

3. $a^2 - 2ab + b^2 = (a - b)^2$

4. $a^3 + b^3 = (a + b)(a^2 - ab + b^2)$

5. $a^3 - b^3 = (a - b)(a^2 + ab + b^2)$

Example 3.4: Factor $4x^2 - 25$.

$$4x^2 - 25 = (2x - 5)(2x + 5) \qquad \text{Use formula 1.}$$

Example 3.5: Factor $x^2 + 6x + 9$.

$$x^2 + 6x + 9 = (x + 3)^2 \qquad \text{Use formula 2.}$$

Example 3.6: Factor $x^2 - 8x + 16$.

$$x^2 - 8x + 16 = (x - 4)^2 \qquad \text{Use formula 3.}$$

Example 3.7: Factor $27x^3 + 8$.

$$27x^3 + 8 = (3x + 2)(9x^2 - 6x + 4) \qquad \text{Use formula 4.}$$

Example 3.8: Factor $x^3 - 125$.

$$x^3 - 125 = (x - 5)(x^2 + 5x + 25) \qquad \text{Use formula 5.}$$

It is possible there will be a need for more than one factoring method to be used in the same problem. It a common factor is present, it is strongly recommended that it be extracted first, then continue to factor whatever remains.

Example 3.9: Factor $2x^2 - 98$ completely.

$$2x^2 - 98 = 2(x^2 - 49) = 2(x - 7)(x + 7)$$

Not every polynomial can be factored. If a polynomial cannot be factored, it is said to be prime.

Example 3.10: Factor $x^2 + 3x + 11$ completely.

This polynomial is prime. No technique above will factor it.

Chapter 4: Rational Expressions and Equations

A rational expression is an expression that involves a polynomial in the numerator and a polynomial in the denominator.

A rational expression is considered simplified when it:

- Is reduced to one fraction.
- Has no common factor in the numerator and the denominator.
- Contains no negative exponents

To simplify rational expressions that are added or subtracted, find a common denominator, if necessary, then add or subtract the numerators as required. (If the denominators differ only in sign, factor a −1 out of one of them.)

Example 4.1: Add $\dfrac{1}{x-4} + \dfrac{x}{4-x}$.

$$\frac{1}{x-4} + \frac{x}{4-x} = \frac{1}{x-4} + \frac{x}{-(x-4)}$$

$$= \frac{1}{x-4} + \frac{-x}{x-4}$$

$$= \frac{1-x}{x-4}$$

Example 4.2: Subtract $\dfrac{7}{x-1} - \dfrac{1}{x^2-3x+2}$

$$\frac{7}{x-1} - \frac{1}{x^2-3x+2} = \frac{7}{x-1} - \frac{1}{(x-2)(x-1)}$$

$$= \frac{7(x-2)}{(x-1)(x-2)} - \frac{1}{(x-1)(x-2)}$$

$$= \frac{7(x-2)-1}{(x-1)(x-2)}$$

$$= \frac{7x-14-1}{(x-1)(x-2)}$$

$$= \frac{7x-15}{(x-1)(x-2)}$$

The only difference in multiplication and division is that in division it is necessary to invert the divisor and multiply.

Example 4.3: Simplify $\dfrac{x^2-25}{x^2+6x+5} \div \dfrac{x-5}{x^2-1}$.

$$\frac{x^2-25}{x^2+6x+5} \div \frac{x-5}{x^2-1} = \frac{(x-5)(x+5)}{(x+1)(x+5)} \div \frac{x-5}{(x-1)(x+1)}$$

$$= \frac{(x-5)(x+5)}{(x+1)(x+5)} \bullet \frac{(x-1)(x+1)}{x-5}$$

$$= \frac{(x-5)(x+5)(x-1)(x+1)}{(x+1)(x+5)(x-5)}$$

$$= x-1$$

Example 4.4: Simplify $\dfrac{\dfrac{1}{x} - \dfrac{1}{y^2}}{\dfrac{1}{xy}}$

$$\frac{\dfrac{1}{x} - \dfrac{1}{y^2}}{\dfrac{1}{xy}} = \left(\frac{1}{x} - \frac{1}{y^2} \right) \div \frac{1}{xy}$$

$$= \left(\frac{y^2}{xy^2} - \frac{x}{xy^2} \right) \div \frac{1}{xy}$$

$$= \left(\frac{y^2 - x}{xy^2} \right) \div \frac{1}{xy}$$

$$= \frac{y^2 - x}{xy^2} \bullet xy$$

$$= \frac{y^2 - x}{y}$$

Solving equations requires the removal of the variable from all denominators. This is done by multiplying through by the least common denominator and canceling. Remember, factoring out a negative sign can keep the common denominator from getting too complex. Also, keep in mind it is never allowed to have a division by zero. Should any apparent answer cause a denominator to be zero, it must be rejected. Not every problem has a solution.

Example 4.5: Solve $\dfrac{4}{x-2} + \dfrac{20}{x+3} = \dfrac{8}{x-2}$, if possible.

The least common denominator is $(x-2)(x+3)$.

$$\frac{4(x-2)(x+3)}{x-2} + \frac{20(x-2)(x+3)}{x+3} = \frac{8(x-2)(x+3)}{x-2}$$

$$4(x+3) + 20(x-2) = 8(x+3)$$

$$4x + 12 + 20x - 40 = 8x + 24$$

$$24x - 28 = 8x + 24$$

$$16x = 52$$

$$x = \frac{52}{16} = \frac{13}{4}$$

Since the solution, $\frac{13}{4}$, is neither -3 nor 2, it is a valid solution.

Example 4.6: Solve $\dfrac{5}{x-1} - \dfrac{1}{1-x} = -4$, if possible.

$$\frac{5}{x-1} - \frac{1}{1-x} = -4$$

$$\frac{5}{x-1} + \frac{1}{x-1} = -4$$

$$\frac{5(x-1)}{x-1} + \frac{x-1}{x-1} = -4(x-1)$$

$$5 + 1 = -4(x-1)$$

$$6 = -4x + 4$$

$$2 = -4x$$

$$-\frac{1}{2} = x$$

A work problem is best thought of as the part of the work done in one hour.

Example 4.7: If Bill can paint a building in 8 hours and Susan can paint the same building in 10 hours, how long would it take if they work together?

Set up the part of the job done in one hour.

$$\frac{1}{8} + \frac{1}{10} = \frac{1}{x}$$

$$\frac{40x}{8} + \frac{40x}{10} = \frac{40x}{x}$$

$$5x + 4x = 40$$

$$9x = 40$$

$$x = \frac{40}{9} \text{ hr}$$

Chapter 5: Radicals and Complex Numbers

Expressions involving radicals are considered simplified when all of the following conditions are met:

- No term contains more than one radical.
- No radical exists in a denominator.
- All perfect squares are removed from a square root, perfect cubes from a cube root, and so on.
- Radicals with powers under them are reduced to the lowest radical and power possible.
- No radical of a radical is allowed.
- Terms with like radicals need to be combined.

Two radical expressions can be multiplied using the same rules as for polynomial multiplication.

Example 5.1: Multiply $(2 + \sqrt{3})(3 - \sqrt{7})$. Simplify the result.

$$(2 + \sqrt{3})(3 - \sqrt{7}) = 2 \bullet 3 - 2\sqrt{7} + 3\sqrt{3} + \sqrt{3 \bullet 7} =$$

$$6 - 2\sqrt{7} + 3\sqrt{3} + \sqrt{21}$$

Example 5.2: Simplify $\dfrac{2}{\sqrt{7}}$

$$\frac{2}{\sqrt{7}} \bullet \frac{\sqrt{7}}{\sqrt{7}}$$

$$= \frac{2\sqrt{7}}{7}$$

In the above example, multiplication by 1 designed to eliminate the radical from the denominator is used.

If the square root in the denominator is part of a two term expression, multiply by 1 in the form of the conjugate over itself. This works because the factoring of $a^2 - b^2$ is $(a+b)(a-b)$, and $a^2 - b^2$ can contain no square roots when a or b contains a square root.

Example 5.3: Simplify $\dfrac{4}{3-\sqrt{5}}$.

$$\frac{4}{3-\sqrt{5}} = \frac{4}{(3-\sqrt{5})}\frac{(3+\sqrt{5})}{(3+\sqrt{5})}$$

$$= \frac{12+4\sqrt{5}}{9-5}$$

$$= \frac{12+4\sqrt{5}}{4} = \frac{4(3+\sqrt{5})}{4}$$

$$= 3+\sqrt{5}$$

Example 5.4: Simplify $\sqrt{2} - \dfrac{3}{\sqrt{2}} + 5\sqrt{8}$.

$$\sqrt{2} - \frac{3}{\sqrt{2}} + 5\sqrt{8} = \sqrt{2} - \frac{3}{\sqrt{2}}\frac{\sqrt{2}}{\sqrt{2}} + 5 \bullet 2\sqrt{2}$$

$$= \sqrt{2} - \frac{3}{2}\sqrt{2} + 10\sqrt{2}$$

$$= \frac{19}{2}\sqrt{2}$$

The even root of an even power gives the absolute value of the variable, because raising to an even power eliminates the sign. Odd roots do not have this condition.

Example 5.5: Simplify $\sqrt{4x^2}$

$$\sqrt{4x^2} = 2|x|$$

Example 5.6: Simplify $\sqrt[3]{16x^3 y^4}$.

$$\sqrt[3]{16x^3 y^4} = 2xy \sqrt[3]{2y}$$

Example 5.7: Simplify $\sqrt[4]{4x^8}$.

$$\sqrt[4]{4x^8} = x^2 \sqrt[4]{4}$$

Notice in the above there is no possibility of x^2 being negative, so the absolute value bars are unnecessary.

Occasionally, working on radicals using the laws of exponents makes things more understandable.

Example 5.8: Simplify $\sqrt[3]{\sqrt{3}}$.

$$\sqrt[3]{\sqrt{3}} = \left((3)^{1/2}\right)^{1/3} = 3^{\frac{1}{2} \cdot \frac{1}{3}} = 3^{1/6} = \sqrt[6]{3}$$

Complex numbers involve a real term and a term containing i or $\sqrt{-1}$. The number i can be squared, and in doing so produces $-1,$ not allowed in the real number system.

Complex umbers should be expressed in the form $a + bi$.

Example 5.9: Simplify $(5 + 2i) - (3 + 6i)$ and express the difference in the form $a + bi.$

$$(5 + 2i) - (3 + 6i) = 5 - 3 + (2 - 6)i$$

$$= 2 - 4i$$

Example 5.10: Simplify $(5 - 2i)(7 + 3i)$ and express the difference in the form $a + bi.$

$$(5 - 2i)(7 + 3i) = 35 + 15i - 14i - 6i^2 = 35 + 15i - 14i - 6(-1)$$

$$= 35 + 6 + 15i - 14i$$

$$= 41 + i$$

Example 5.11: Simplify $\dfrac{9 - i}{4 + 3i}$ and express the difference in the form $a + bi.$

$$\frac{(9-i)}{(4+3i)} \cdot \frac{(4-3i)}{(4-3i)} = \frac{36-27i-4i+3i^2}{16-9i^2}$$

$$= \frac{36-27i-4i-3}{16+9}$$

$$= \frac{33-31i}{25} \text{ or } \frac{33}{25} + \frac{-31}{25}i$$

Chapter 6: Quadratic Equations

A quadratic equation has the form, or can be put in the form,

$$ax^2 + bx + c = 0.$$

An extremely powerful property of the number zero is that if he product of two numbers is zero, at least one of the two numbers is zero. If an expression can be factored, the two solutions come from setting the factors equal to zero.

Example 6.1: Solve the equation $x^2 - 10x = -24$.

First, put the equation in the proper form, $x^2 - 10x + 24 = 0.$

Next, factor. $(x - 6)(x - 4) = 0$

Set each factor equal to zero. $x - 6 = 0$ and $x - 4 = 0$

Solve the linear equations that result. $x = 6$ and $x = 4$

Not every quadratic equation is suitable for factoring. If a quadratic equation cannot be factored, another technique is to complete the square, then take both square roots and solve. This can be avoided by completing the square once using a general quadratic equation, then simply plugging into the quadratic formula that results.

The development, using completion of the square, of the quadratic formula is now given.

$$ax^2 + bx + c = 0$$

$$ax^2 + bx = -c$$

$$x^2 + \frac{b}{a}x = -\frac{c}{a}$$

$$x^2 + \frac{b}{a}x + \frac{b^2}{4a^2} = \frac{b^2}{4a^2} - \frac{c}{a}$$

$$\left(x + \frac{b}{2a}\right)^2 = \frac{b^2 - 4ac}{4a^2}$$

$$x + \frac{b}{2a} = \pm\sqrt{\frac{b^2 - 4ac}{4a^2}}$$

$$x = -\frac{b}{2a} \pm \frac{\sqrt{b^2 - ac}}{2a}$$

$$x = \frac{-b \pm \sqrt{b^2 - 4ac}}{2a}$$

Example 6.2: Solve $x^2 - 3x + 6 = 0$.

$$x = \frac{3 \pm \sqrt{9 - 24}}{2} = \frac{3 \pm \sqrt{-15}}{2}$$

$$= \frac{3 \pm i\sqrt{15}}{2}$$

Example 6.3: An object is thrown upward at 64 ft/s, and eventually comes back to the ground. The time in seconds, *t*, the object is in the air to be at a given height in feet, *h*, is given by the formula $h = -16t^2 + 64t$. How long will it take to come back down?

$$h = -16t^2 + 64t$$

The ground is at $h = 0$, $\quad 0 = -16t^2 + 64t$

Solve for *t* by factoring. $\quad 0 = -16t(t - 4)$

$t = 0$ and $t - 4 = 0$

So, $t = 4$

Reject $t = 0$, since that is when the object is thrown, not when it comes back down. The object comes down after 4 seconds.

Chapter 7: Radical Equations

When an equation contains one or more radicals, the technique is to first isolate one radical term on one side of the equation. Each side of the equation is then raised to the appropriate power to eliminate the radical. This is to be repeated until all radicals have been removed. The resulting equation is to be solved by known methods. Finally, the solution(s) should be checked if the equation was raised to an even power, since there may be solutions to the new equation that do not solve the original equation.

Example 7.1: Solve the equation $\sqrt{3x-2} = 2x-1$.

$$\sqrt{3x-2} = 2x-1$$

$$\left(\sqrt{3x-2}\right)^2 = (2x-1)^2$$

$$3x-2 = 4x^2 - 4x + 1$$

$$0 = 4x^2 - 7x + 3$$

$$0 = (4x-3)(x-1)$$

$$4x-3 = 0 \text{ and } x-1 = 0$$

$$x = \frac{3}{4} \text{ and } x = 1$$

Check: $\sqrt{\dfrac{9}{4}} - 2 = \dfrac{6}{4} - 1$ $\dfrac{1}{2} = \dfrac{1}{2}$, TRUE

$\sqrt{3(1) - 2} = 2(1) - 1$ 1 = 1, TRUE

The solutions are $x = 1, \dfrac{3}{4}$.

Example 7.2: Solve the equation $\sqrt{x - 4} = \sqrt{x + 4} - 2.$

$\sqrt{x - 4} = \sqrt{x + 4} - 2$

$\left(\sqrt{x - 4}\right)^2 = \left(\sqrt{x + 4} - 2\right)^2$

$x - 4 = x + 4 - 4\sqrt{x + 4} + 4$

$-12 = -4\sqrt{x + 4}$

$12^2 = \left(-4\sqrt{x + 4}\right)^2$

$144 = 16(x + 4)$

$9 = x + 4$

$5 = x$

Check: $\sqrt{5-4} = \sqrt{5+4} - 2$

$1 = 1$ TRUE

The solution is $x = 1$.

Chapter 8: Straight Lines

It takes two points, or one point and a slope, to define a straight line. The equation for a slope is $m = \dfrac{y_2 - y_1}{x_2 - x_1}$, where two points are used.

Example 8.1: Find the slope of the line through (4, 2) and (-5, 3).

Using the points given, $m = \dfrac{3 - 2}{-5 - 4} = -\dfrac{1}{9}$.

The slope is -1/9.

This information can be used to determine the equation of a straight line. In using the point-slope formula, $y - y_1 = m(x - x_1)$, either point may be used.

Example 8.2: Find the equation of the line between (6, 8) and (3, 9).

The slope is $m = \dfrac{9 - 8}{3 - 6} = -\dfrac{1}{3}$

Using the formula, $y - 8 = -\dfrac{1}{3}(x - 6)$.

$$y - 8 = -\dfrac{1}{3}x + 2$$

$$y = -\frac{1}{3}x + 10$$

It is easy to extract information from the equation for a straight line. The slope-intercept form, $y = mx + b$, is most useful for this purpose. Simply take the coefficient of x as the slope, and the constant as the y-value of the y-intercept.

Example 8.3: What is the slope and what is the y-intercept of the straight line defined by

$$3x - 5y = 30 ?$$

First, solve for y.

$$-5y = -3x + 30$$

$$y = \frac{3}{5}x - 6$$

The slope is $\frac{3}{5}$, and the y-intercept is -6. (More properly, the intercept is a point and should be expressed $(0, -6)$.

Two lines are parallel if and only if they either have the same slope, or if they are both vertical lines. The lines are perpendicular if and only if their slopes are negative reciprocals of each other, or if one is horizontal and the other is vertical.

Example 8.4: What is the equation of the line parallel to $y = 3x + 6$ that passes through

the point $(5, -1) ?$

The slope is 3, so the equation is $y + 1 = 3(x - 5)$.

$$y = 3x - 16$$

Example 8.5: What is the equation of the line perpendicular to $y = 3x + 6$ that passes through the point $(6, -1)$?

The slope of 3 has a negative reciprocal of $-\dfrac{1}{3}$. $\quad y + 1 = -\dfrac{1}{3}(x - 6)$

$$y = -\frac{1}{3}x + 1$$

     To graph a straight line, simply plot two points and connect them. To graph a linear inequality, plot the straight line associated with the boundary of the inequality, and pick one point not on the line. Note the line is solid if the equality is included, and dashed if not. The point chosen should be used in the original inequality, and if it makes a true statement, shade the same side of the boundary as the point, if not, shade he other side.

Chapter 9: Systems of Equations

It is possible to solve a system of equations be the substitution method, or by the elimination (sometimes called the addition) method.

Example 9.1: Solve the system of equations, or state no solution exists.

$$x + 3y = 11$$
$$2x - y = 1$$

Using the substitution method, solve one equation for one variable, and substitute the resulting expression into the other expression.

$$x = 11 - 3y$$
$$2(11 - 3y) - y = 1$$
$$22 - 6y - y = 1$$
$$22 - 7y = 1$$
$$-7y = -21$$
$$y = 3$$

Substituting and solving for x, $x = 11 - 3(3)$.

$$x = 11 - 9 = 2$$

The solution is $(2, 3)$.

Using the elimination method, multiply one or both equations so as to eliminate a variable when the two equations are added.

Multiply equation 2 by 3

$$x + 3y = 11$$
$$6x - 3y = 3$$

$$7x = 14.$$
$$x = 2$$

$$2 + 3y = 11$$
$$3y = 9$$
$$y = 3$$

The solution is $(2, 3)$.

If both variables disappear, and the resulting statement is true, there are dependant solutions. However, if the resulting equation is false, there is no solution.

Example 9.2: Solve the system of equations, or state no solution exists.

$$4x + 5y = 2$$
$$8x + 10y = 12$$

Multiply the first equation by -2.

$$-8x - 10y = -4$$
$$8x + 10y = 12$$

$0 + 0 = 8$

No solution.

Example 9.3: Solve the system of equations, or state no solution exists.

$$3x - 6y = -6$$
$$-x + 2y = 2$$

Multiply the second equation by 3.

$$3x - 6y = -6$$
$$-3x + 6y = 6$$

$0 + 0 = 0$

There are an infinite number of solutions such that if one variable is picked, the other is forced. To display these solutions in parametric form, choose one variable and represent it with a parameter, then solve for the other variable in terms of the parameter.

Let $x = t$, any real number. Then $3t - 6y = -6$.

$$y = \frac{3t}{2} + 1$$

The solutions are given by $\left(t, \frac{3t}{2} + 1 \right)$, t any real number.

Chapter 10: Functions

A *function* relates an element of the *domain* (the allowed values for which the function is defined) to an element in the *range* (the result of applying the function rule to the domain elements).

Example 10.1: Evaluate $f(x) = 3x - 6$, when x is 4.

$$f(4) = 3(4) - 6 = 12 - 6 = 6$$

Example 10.2: What is the domain and what is the range of

$$f(x) = \sqrt{x-2} + 8?$$

The domain consists of all allowable values of x, which in this case are those values that do not produce a negative value under the radical symbol.

The domain is $x \geq 2$. $[2, \infty)$.

The range consists of all possible $f(x)$ values. The range is $[8, \infty)$.

Example 10.3: What is the domain and range of $f(x) = \dfrac{1}{x}$

Since division by 0 is not allowed, the domain is all x except 0.

The graph shows the function never becomes 0, so the range is also all values except 0.

Composite functions, $f(g(x))$, also represented by $f \circ g(x)$, is the function form of f with the variable replaced with the expression for $g(x)$.

Example 10.4: If $f(x) = 3x^2 + 4$ and $g(x) = 2x^3$ find $f(g(x))$

$$f(g(x)) = 3(2x^3)^2 + 4 = 12x^6 + 4$$